# INTERNATIONAL CENTRE FOR MECHANICAL SCIENCES

COURSES AND LECTURES - No. 170

VICTOR SZEBEHELY
UNIVERSITY OF TEXAS, AUSTIN

# THE GENERAL AND RESTRICTED PROBLEMS OF THREE BODIES

COURSE HELD AT THE DEPARTMENT
OF GENERAL MECHANICS
SEPTEMBER 1973

UDINE 1974

SPRINGER-VERLAG WIEN GMBH

This work is subject to copyright.

All rights are reserved,

whether the whole or part of the material is concerned

specifically those of translation, reprinting, re-use of illustrations,

broadcasting, reproduction by photocopying machine

or similar means, and storage in data banks.

© 1972 by Springer-Verlag Wien

Originally published by Springer-Verlag Wien New York in 1972

ISBN 978-3-211-81264-8  ISBN 978-3-7091-2916-6 (eBook)
DOI 10.1007/978-3-7091-2916-6

# PREFACE

*The restricted problem of three bodies is introduced as a special form of the general gravitational three-body problem. The equations of motion are derived in a fixed (sidereal) and in a rotating (synodic) coordinate system. From the latter the Jacobian integral immediately follows. Its applications to establish limiting regions of possible motions are discussed in detail. While in the restricted problem the masses of two of the participating bodies are finite and the third body is of infinitesimal mass, in the general problem all masses are of the same order of magnitude. Therefore, planetary and lunar theories belong to the restricted problem. The equations of motion of the general problem are given and the Jacobian form is derived. The Lagrange-Jacobi equation, combined with Sundman's inequality allows the presentation of the most recent results in this field. While emphasis in the restricted problem is put on the periodic solutions, the general problem is based on asymptotic escape solutions.*

Victor Szebehely

Udine, September 1973

# INTRODUCTION

These lectures present an approach to celestial mechanics which combines qualitative and quantitative investigations. Several methods and tools are given and it is shown how some of these are applied to the problem of three bodies, which is considered the simplest nontrivial question of interest in celestial mechanics.

The audience is acquainted first with some of the important aspects of the restricted problem of three bodies. Starting with the fourth order system of differential equations of motion, several representations of the totality of solutions are given.

These lectures use George Birkhoff's celebrated 1915 paper on the restricted problem and his book entitled "Dynamical Systems". It is the ambition of this writer to bring Birkhoff's brilliant contributions to the attention of the audience. It might take a century to truly appreciate his and Poincaré's contributions to dynamics and to thoroughly understand and apply their ideas. If the audience will enjoy the fresh breeze of "rationale", if they will succeed in communing with Poincaré's spirit, if they will admire the challenging depth of Birkhoff's ideas, and the century proclaimed above herewith is shortened, then the lamplighter activities of these lectures are not in vain.

From an advanced point of view, the aim of dynamics is to characterize completely the totality of the possible motions of dynamical systems by their qualitative properties. The recognition of this fact is the result of the following historical development of theoretical dynamics :

1. Sir Isaac Newton and his contemporaries, during the initial phase of development, attempted to find explicit expressions for coordinates representing the motion of dynamical systems as functions of time.

2. Euler and Laplace emphasized series solutions and successive approximation techniques during the second phase.

3. The third phase was associated with the Lagrange-Hamilton-Jacobi school which considered dynamics an exercise in variational and maximum-minimum problems.

4. Progression from the formal and computational phases to a phase where qualitative matters dominate was made by the work of Poincaré and Birkhoff. This school justifies one of Poincaré's (*) salse dicta : "Les mathématiciens n'étudient pas des objects, mais des relations entre les objects ; il leur est donc indifférent de remplacer ces objects par d'autres, pourvu que les relations ne changent pas. La matière ne leur importe pas, la forme seule les intéresse".

---

(*) This writer wishes to pay tribute to Poincaré, who included paragraphs from Hill's in English in his celebrated Méthodes Nouvelles.

# Qualitative Dynamics

The topological approach in dynamics is a good example for the above quotation. In fact, it might even be considered a necessary approach if it is recognized that many existence proofs of solutions of differential equations are often of local character. Many problems and fundamental questions in dynamics are, on the other hand, problems "in the large" and are controlled by the structure of the nonlocal topology of the manifold involved.

Birkhoff's and Poincaré's work fits admiringly well into the developmental picture of dynamics. It has been shown by Klose that their work is also significant in celestial mechanics by showing that the restricted three-body problem might furnish information essential in lunar (satellite) and planetary problems. The fact that bodies in the solar system might be arranged according to their masses in three groups (sun, planets, and everything else) shows immediately the extended range of possibilities regarding the applications of the restricted three-body problem to dynamical astronomy.

The following five special "qualitative" areas of analytical dynamics might be of interest:

1. Modes of motion of dynamical systems in which the same absolute or relative configuration of the system is repeated at regular intervals of time are called <u>periodic motions</u>. The existence of periodic motions has been studied along four major lines :

A. The method of analytical continuation which is based on Cauchy's local existence theorems is associated with Poincaré.

B. The geodesic interpretation of Hadamard, Poincaré, Whittaker and Kirkhoff.

C. The topological methods expanded by Poincaré and Birkhoff.

D. The method of comparing undetermined Fourier coefficients. Hill's name is associated with this technique; however, it is to be recalled that he has not shown convergence for his series and as a consequence did not give an existence proof. (This was done by Liapounoff in 1895 and by Witner in 1925).

The question regarding the paramount importance of periodic solutions in the restricted problem and why such solutions have received and should receive attention might be answered by realizing that other solutions present very difficult problems. This reasoning by default can be substantiated by Poincaré's conjecture according to which the distribution of periodic solutions in the restricted problem is dense, i.e. close to any motion, periodic motions can be found which can be used to build other solutions. Considering also the well known difficulties encountered in celestial mechanics regarding the convergence problem of trigonometric series obtained by perturbation techniques, one welcomes the existence of periodic solutions, since this will assure the convergence of trigonometric series representing such periodic motions. The present state of art suggests that the main reason for studying such motions is that a number of "natural phenomena show periodicity in some sense" (Birkhoff) and that the examination of the existing literature shows very little progress in connection with motions of more general type.

2. The second large qualitative problem area, **integrability**, is mostly a matter of definition. According to one definition, a system is integrable if all the coordinates can be expressed as "known functions" of time. The set of "known functions" is, of course, varying ; that is the set itself is a function of time ! Painlevé, in his Stockholm lectures (1895) for instance, admitted all functions to the set of "known functions" which are defined by uniformly convergent infinite series. According to this, then, the restricted three-body problem becomes integrable since singularities of collisions can be eliminated by change of the variables. We must remark, however, that this type of integrability is of doubtful importance and it is well known that the Sundman series are of limited value for either computations or for obtaining qualitative information. It is noted that Sundman's results were given without proof prior to him by Bruns and Weirstrass. Astronomers were skeptical of the usefulness of the Sundman series from the very beginning, and even the elegant approach given by Levi-Civita using canonical variables didn't raise the value of the Sundman series.

Another definition of integrability of a dynamical system could be associated with the possibility of reducing the differential equations to quadratures. It can be shown, however, that this is neither a necessary nor a sufficient condition for obtaining qualitative information about dynamical systems.

The best characterization of the integrability problem is probably obtained by referring once again to Poincaré's somewhat flippant remark according to which a system is neither integrable nor nonintegrable, but more or less integrable.

3. Much can be said about the <u>reducibility</u> of the problem. By reducibility we don't necessarily mean the reduction of the order of the system of differential equations under consideration. In a more general sense, reducibility means the reduction of a problem, or the changing of the problem, into another one. In the restricted three-body problem, the originally fourth order system of differential equations together with the Jacobian integral suggests that the problem can be presented in a third order form. This, in turn, suggests the analogy to the steady flow of an incompressible fluid in which certain streamlines correspond to periodic orbits. Poincaré has shown that the streamline picture and its property of repeatedly cutting a stationary surface is identical to the transformation of the surface into itself. In this way he has demonstrated that the re-

stricted three-body problem can be reduced to a problem in surface transformations.

4. **Stability** can be often demonstrated by an inequality restricting the coordinates of a dynamical system. One of the interesting results contributed to Hill shows that the Jacobian integral restricts the moon's motion to a certain region around the earth.

5. The fifth group of "qualitative" examples might be furnished by <u>existence problems</u> associated with certain specific properties and by problems connected with large values of the independent (time) variable $(t \to \infty)$. Regarding the second problem, Poincaré has shown that the series used in lunar and similar theories are divergent in general but are suitable for calculations and that these series represent the coordinates in an asymptotic sense. This negative result certainly does not take care of the $t \to \infty$ problems in general but it shows that these problems are some times amenable to either numerical or analytical approaches.

The above five groups of examples are ample justifications for the need for qualitative studies. These studies often lead to disappointments since no new computational methods and no numerical miracles are presented, no orbits are computed, and not even differential equations are being solved. Nevertheless, it is expected that the qualitative approach will answer some of the questions reviewed above regarding periodic motions, integrability, reducibility, stability, existence, etc..

Questions and problems will be mentioned as possibilities for further research in areas where the tools presented can be utilized. For instance, the so-called satellite problem of the restricted problem of three bodies will be discussed in some detail, while the so-called planet problem and the comet problem will only be referred to.

It seems to be eminently cleat that joint <u>qualitative</u> and <u>quantitative</u> investigations of "nonintegrable dynamical problems" are essential for the further progress in dynamics and also in celestial mechanics.

## THE RESTRICTED PROBLEM
### Statement of the Problem and Equations of Motion

In the restricted circular planar problem of three bodies, two bodies (assumed to be point masses and calles primaries) revolve around their center of mass in circular orbits under the influence of their mutual gravitational attraction. A third body (attracted by the previous two but not influencing their motion) moves in the plane defined by the two revolving bodies. The problem is to determine the motion of this third body, (Szebehely, 1967).

Let the masses of the primaries be $m_1$ and $m_2$, their mean motion $n$, and their distance $\ell$ (see Figure 1).

Then

$$k^2 M = n^2 \ell^2$$

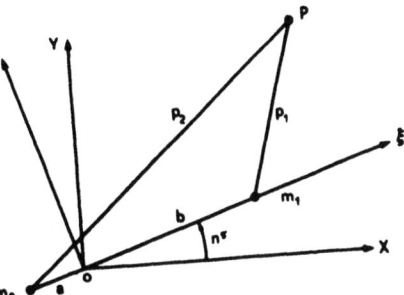

Fig. 1. Planar Circular Restricted Problem in Fixed (X,Y) and Rotating ($\xi,\eta$) Coordinate Systems

where $M = m_1 + m_2$.

The center of mass of the system is located on the line connecting $m_1$ and $m_2$, and its distance from $m_2$ and $m_1$, is respectively

$$a = \frac{m_1 \ell}{M} \quad , \quad b = \frac{m_1 \ell}{M} \quad , \quad \ell = a + b \, .$$

Taking the origin of a fixed inertial coordinate system (X,Y) at the mass center (O), and using $\tau$ for time, the equations of motion referred to this system will be

$$\frac{d^2 X}{d\tau^2} = \frac{\partial F}{\partial X} \quad , \quad \frac{d^2 Y}{d\tau^2} = \frac{\partial F}{\partial Y} \, ,$$

where F is Poincaré's "force function" given by

$$F = k^2 \left( \frac{m_1}{\rho_1} + \frac{m_2}{\rho_2} \right)$$

with $\rho_1$ and $\rho_2$ being the distances between the primaries and the third body.

Introducing a uniformly rotating coordinate system ($\xi,\eta$) with origin at the mass center so that $m_1$ and $m_2$ are located on the $\xi$ axis with coordinates (b,0) and (-a,0), the equations of motion become

$$\frac{d^2\xi}{d\tau^2} - 2n\frac{d\eta}{d\tau} = \frac{\partial F^*}{\partial \xi},$$

$$\frac{d^2\eta}{d\tau^2} + 2n\frac{d\xi}{d\tau} = \frac{\partial F^*}{\partial \eta},$$

where
$$F^* = F + \frac{1}{2}n^2(\xi^2 + \eta^2)$$

and
$$\rho_1^2 = (\xi - b)^2 + \eta^2, \quad \rho_2^2 = (\xi + a)^2 + \eta^2$$

The introduction of nondimensional quantities simplifies the equations. Let

$$x = \xi/\ell, \quad y = \eta/\ell, \quad r_1 = \rho_1/\ell, \quad r_2 = \rho_2/\ell, \quad t = n\tau,$$

$$\mu = \frac{m_2}{M}.$$

The equations of motions in nondimensional form are:

$$\frac{d^2 x}{dt^2} - \frac{dy}{dt} = \frac{\partial \bar{\Omega}}{\partial x},$$

$$\frac{d^2 y}{dt^2} + 2\frac{dx}{dt} = \frac{\partial \bar{\Omega}}{\partial y},$$

where
$$\bar{\Omega} = \frac{1}{2}(x^2 + y^2) + \frac{\mu}{r_2} + \frac{1-\mu}{r_1} = \frac{F^*}{\ell^2 n^2}$$

and
$$r_1^2 = (x - \mu)^2 + y^2, \quad r_2^2 = (x + 1 - \mu)^2 + y^2.$$

The above four equations represent the problem in conventional nondimensional quantities. The corresponding physic al picture is shown on Figure 2.

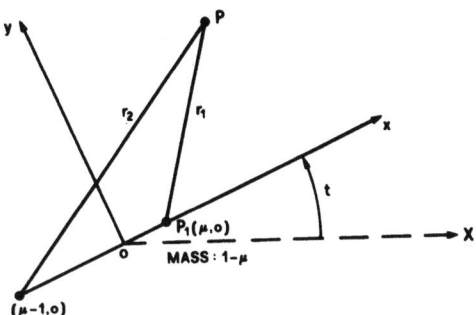

Fig. 2. Planar Circular Restricted Problem in Nondimensional Rotating System.

The Jacobian integral is obtained by multiplying the equations of motion by $2\,dx/dt$, by $2\,dy/dt$, adding and integrating :

$$\left(\frac{dx}{dt}\right)^2 + \left(\frac{dy}{dt}\right)^2 = 2\bar{\Omega} - \bar{C},$$

where $\bar{C}$ is the constant of integration.

A symmetrical form of $\bar{\Omega}$ is obtained by adding the constant quantity :

$$\frac{1}{2}\,\mu\,(1-\mu)$$

and introducing in this way

$$\Omega = \frac{1}{2}\left[(1-\mu)\,r_1^2 + \mu\,r_2^2\right] + \frac{1-\mu}{r_1} + \frac{\mu}{r_2}.$$

The new Jacobian constant C is related to the previous one by $C = \bar{C} + \mu(1-\mu)$.

## The Problem of Three Bodies and the Restricted Problem

The problem of three bodies is defined as follows : three particles attract each other according to the Newtonian law of gravitation, they are free to move in space and are initially moving in any given manner ; find their subsequent motion.

Figure 3 shows the three bodies and the corresponding vectors : $\bar{r}_1$, $\bar{r}_2$, and $\bar{r}_3$. The coordinates of these vectors are denoted by $q_i$ and $r_1(q_1, q_2, q_3)$, $r_2(q_4, q_5, q_6)$, $r_3(q_7, q_8, q_9)$.

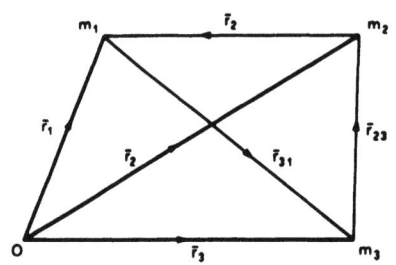

Fig. 3. The problem of three bodies: the masses are $m_1$, $m_2$, and $m_3$; the corresponding position vectors are $\bar{r}_1, \bar{r}_2$ and $\bar{r}_3$.

The vectors connecting the mass points are

$$\bar{r}_{12} = \bar{r}_1 - \bar{r}_2, \quad \bar{r}_{23} = \bar{r}_2 - \bar{r}_3, \quad \bar{r}_{31} = \bar{r}_3 - \bar{r}_1,$$

and the distances between $m_1$, $m_2$, and $m_3$ are

$$\bar{r}_{12} = \left[(q_1 - q_4)^2 + (q_2 - q_5)^2 + (q_3 - q_6)^2\right]^{1/2},$$

$$\bar{r}_{23} = \left[(q_4 - q_7)^2 + (q_5 - q_8)^2 + (q_6 - q_9)^2\right]^{1/2},$$

$$\bar{r}_{31} = \left[(q_7 - q_1)^2 + (q_8 - q_2)^2 + (q_9 - q_3)^2\right]^{1/2}.$$

# The General Problem of Three Bodies

The kinetic energy of the system is

$$T = \frac{1}{2} m_1(\dot{q}_1^2 + \dot{q}_2^2 + \dot{q}_3^2) + \frac{1}{2} m_2(\dot{q}_4^2 + \dot{q}_5^2 + \dot{q}_6^2) + \frac{1}{2} m_3(\dot{q}_7^2 + \dot{q}_8^2 + \dot{q}_9^2).$$

The momenta are given by

$$p_i = \frac{\partial T}{\partial \dot{q}_i}, \quad (i = 1, \ldots 9)$$

or

$$\begin{pmatrix} p_1 \\ p_2 \\ p_3 \end{pmatrix} = m_1 \begin{pmatrix} \dot{q}_1 \\ \dot{q}_2 \\ \dot{q}_3 \end{pmatrix}, \begin{pmatrix} \dot{q}_4 \\ \dot{q}_5 \\ \dot{q}_6 \end{pmatrix} = m_2 \begin{pmatrix} \dot{q}_4 \\ \dot{q}_5 \\ \dot{q}_6 \end{pmatrix}, \begin{pmatrix} \dot{q}_7 \\ \dot{q}_8 \\ \dot{q}_9 \end{pmatrix} = m_3 \begin{pmatrix} \dot{q}_7 \\ \dot{q}_8 \\ \dot{q}_9 \end{pmatrix},$$

or simply by

$$p_i = m_k \dot{q}_i,$$

where $k$ is the integer part of $i + 2/3$, i.e.,

if $i = 1, 2,$ and $3$     then $k = 1$,
if $i = 4, 5,$ and $6$     then $k = 2$,
if $i = 7, 8,$ and $9$     then $k = 3$.

The Hamiltonian of the system is

$$H = \frac{1}{2} \sum_{i=1}^{9} p_i \dot{q}_i - L,$$

where L is the Lagrangian function,

$$L = T - V$$

and V is the potential. Eliminating the velocities from the

Hamiltonian, we have

$$H = \frac{1}{2} \sum_{i=1}^{9} \frac{p_i^2}{m_i} + V,$$

and the equations of motion become

$$\dot{q}_i = \frac{\partial H}{\partial p_i} \quad \text{and} \quad \dot{p}_i = -\frac{\partial H}{\partial q_i},$$

with $i = 1 \ldots 9$.

The potential function of the system is

$$V = -G \left( \frac{m_1 m_2}{|\bar{r}_{12}|} + \frac{m_2 m_3}{|\bar{r}_{23}|} + \frac{m_3 m_1}{|\bar{r}_{31}|} \right) = -F,$$

where the gravitational constant has the value of unity.

The system possesses 9 degrees of freedom, since the three bodies have 3 position coordinates each. Lagrange showed that this 18-th order system of differential equations can be reduced to a 6th order system (1772). The reduction is performed as follows:

1. Since no external forces are acting on the system the center of mass will move on a straight line with constant velocity, i.e.

$$\sum_{i=1}^{3} m_i \ddot{\bar{r}}_i = \bar{a} \quad \text{and} \quad \sum_{i=1}^{3} m_i \dot{\bar{r}}_i = \bar{a}t + \bar{b}$$

These two vector equations, corresponding to six scalar equations with six scalar constants of integration (i.e. the coordinates of $\bar{a}$ and $\bar{b}$) represent six integrals with which

# Integrals of the General Problem

the system can be reduced to the $(18 - 6 =)$ 12th order.

2. The conservation of angular momentum can be written as

$$\sum_{i=1}^{3} \bar{r}_i \times m_i \dot{\bar{r}}_i = \bar{c}$$

and the corresponding three scalar integrals (with the components of $\bar{c}$ as integration constants) is used to reduce the system from the 12th to the 9th order.

3. In addition to the above 9 integrals, the energy conservation principle will furnish the 10th and last integral, by means of which the order can be reduced to 8. Further reduction is possible by the process called the "elimination of the nodes" and by eliminating the time.

With the 10 integrals and the two elimination processes the original 18th order system can be reduced to an $(18 - 10 - 2 =)$ 6th order system, as stated above. The actual process of reduction is a major task which will not be given here. The reader is referred, for instance, to Whittaker's (1944) book.

The equations of motion will be given now in an explicit form so that the corresponding equations for the restricted problem of three bodies may be introduced. The equations of motion may be written as :

$$m_i \ddot{\bar{r}} = -\frac{\partial V}{\partial \bar{r}_i} \quad , \quad i = 1, 2, 3 ,$$

where the right hand sides represent the gradients of the potential. These will now be evaluated. First we note that

$$\frac{\partial V}{\partial r_1} = \frac{\partial}{\partial r_1}\left(\frac{m_1 m_2}{|\bar{r}_{12}|} + \frac{m_3 m_1}{|\bar{r}_{31}|}\right),$$

and that

$$\frac{\partial}{\partial \bar{r}_1}|\bar{r}| = \frac{\bar{r}}{|\bar{r}|}.$$

Consequently

$$\frac{\partial}{\partial \bar{r}_1}\frac{1}{|\bar{r}_{12}|} = -\frac{\bar{r}_1 - \bar{r}_2}{|\bar{r}_1 - \bar{r}_2|^3},$$

and

$$\frac{\partial}{\partial \bar{r}_1}\frac{1}{|\bar{r}_{31}|} \quad \frac{\bar{r}_3 - \bar{r}_1}{|\bar{r}_3 - \bar{r}_1|^3}.$$

The other four remaining gradients may be obtained the same way.

The equations of motion become

$$\ddot{\bar{r}}_1 = -m_2\frac{r_1 - r_2}{|r_1 - r_2|^3} + m_3\frac{r_3 - r_1}{|r_3 - r_1|^3},$$

$$\ddot{\bar{r}}_2 = -m_3\frac{\bar{r}_2 - \bar{r}_3}{|\bar{r}_2 - \bar{r}_3|^3} + m_1\frac{\bar{r}_1 - \bar{r}_2}{|\bar{r}_1 - \bar{r}_2|^3},$$

# Equations of Motion of the General Problem

$$\ddot{\bar{r}}_3 = -m_1 G \frac{\bar{r}_3 - \bar{r}_1}{|\bar{r}_3 - \bar{r}_1|^3} + m_2 G \frac{\bar{r}_2 - \bar{r}_3}{|\bar{r}_2 - \bar{r}_3|^3} \;.$$

where the first equation was divided by $m_1$, the second by $m_2$, and the third by $m_3$.

These equations describe the general case of the problem of three bodies with Newtonian gravitational forces. Their structure is of some interest in as much as the masses, $m_1$, $m_2$, and $m_3$ are missing from the first, second, and third equation respectively. (This fact, of course, does not "uncouple" the equations since all three position vectors occur in all three equations).

The terms appearing on the right hand side have specific physical significance. The first term on the right side of the first equation, for instance, represents the force per unit mass acting on the first body due to the presence of the second mass. The second term on the right side of the same equation is the effect of the third mass $(m_3)$ on $m_1$.

Decreasing the mass of the third body will reduce its influence on the motion of $m_1$ and $m_2$, i.e. as $m_3 \rightarrow 0$ the first two equations of motion become

$$\ddot{\bar{r}}_1 = -m_2 G \frac{\bar{r}_1 - \bar{r}_2}{|\bar{r}_1 - \bar{r}_2|^3} \;,$$

$$\ddot{\bar{r}}_2 = -m_1 G \frac{\bar{r}_2 - \bar{r}_1}{|\bar{r}_2 - \bar{r}_1|^3} \quad ,$$

while the third equation will not change. This step <u>does uncouple</u> the equations since the motion of $m_1$ and $m_2$ can now be determined without considering the effect of the third mass by solving the above 12th order system of differential equations.

The third original equation of motion requires comment since if in the first two equations $m_3 = 0$, then the third equation reduces to $0 = 0$ on account of the fact that in this case no division by $m_3$ is allowed. Herein lies the approximation which creates the restricted problem of three bodies. The assumption is made that $m_3 \ne 0$ but that $m_3$ is sufficiently small so that it does not effect the motion of $m_1$ and of $m_2$. That is, the above two equations are approximate while the third equation of motion is exact. The system of equations consisting of the above two and of the third original equation of motion represent our dynamical system only approximately; the degree of approximation is given by the "smallness" of the terms

$$m_3 \frac{\bar{r}_3 - \bar{r}_1}{|\bar{r}_3 - \bar{r}_1|^3} \quad \text{and} \quad m_3 \frac{\bar{r}_2 - \bar{r}_3}{|\bar{r}_2 - \bar{r}_3|} \quad ,$$

as compared to the terms

$$m_2 \frac{\bar{r}_1 - \bar{r}_2}{|\bar{r}_1 - \bar{r}_2|^3} \quad \text{and} \quad m_1 \frac{\bar{r}_1 - \bar{r}_2}{|\bar{r}_1 - \bar{r}_2|^3} \quad .$$

# Reduction to the Restricted Problem

The effect of $m_1$ and $m_2$ on the motion of $m_3$ is given by the third original equation of motion. Accepting the above-mentioned approximation one might solve the above two equations, substitute the solution into the third equation of motion and obtain the sixth order differential equation for the motion of the third body :

$$\ddot{\bar{r}}_3 = - m_1 G \frac{\bar{r}_3 - \bar{r}_1}{|\bar{r}_3 - \bar{r}_1|^3} + m_2 G \frac{\bar{r}_2 - \bar{r}_3}{|\bar{r}_2 - \bar{r}_3|^3} \ ,$$

where now $\bar{r}_1$ and $\bar{r}_2$ are given functions of the time and of the initial conditions.

This equation describes the restricted problem of three bodies. The "restriction" is equivalent to the assumption according to which the motion of $m_1$, and $m_2$ is not influenced by $m_3$, while the motion of $m_3$ is determined by the masses and by the motion of $m_1$ and $m_2$.

The last equation can be generalized to

$$\ddot{\bar{r}}_3 = \bar{f}(m_1, m_2, \bar{r}_1, \bar{r}_2, \bar{r}_3) \ ,$$

where $\bar{r}_3$ is the only unknown function of time, $m_1$, and $m_2$ are given constant and $\bar{r}_1$ and $\bar{r}_2$ are also given as functions of the time and of their initial conditions. The function $\bar{f}$ in this equation represents the force field which usually is the Newtonian gravitational field. In addition to this equation the initial conditions of the third body will be needed, in order to determine its motion.

The classification of the various forms of the restricted problem follows from the above remarks and is based on the above equation.

1. Depending on the force law ($\bar{f}$) we speak of Newtonian and non-Newtonian restricted problems.

2. Depending on the initial conditions for $\bar{r}_1$ and $\bar{r}_2$, in the Newtonian case we speak of the circular of the general conic section restricted problems.

3. Depending on the initial conditions of the third body we distinguish between the planar and the three dimensional restricted problems. In a Newtonian gravitational field the first two bodies (primaries or principal bodies) will always move in a plane and if the third body's initial conditions are such that its initial velocity vector is in the plane determined by the orbits of the primaries, it will stay in this plane.

4. Further classification is possible by specifying the $m_2/m_1$ ratio. The Coppenhagen restricted problem, for instance, is distinguished by using unity for this ratio.

More attention has been devoted to the Newtonian, circular, planar restricted problem than to any other variations, however, the Newtonian elliptic planar problem and the Newtonian circular three dimensional problem have also been studied.

Several qualitative and analytical results are

available without numerical specification of the $m_2/m_1$ ratio. Quantitative treatments of course require a certain numerical value and the 1:1, 1:10, 1:80 and 1:1000 approximate ratios dominate, corresponding to the Coppenhagen problem, to G. Darwin's work, to the Moon-Earth problem and to the Jupiter-Sum problem.

### Regions of Possible Motion

An important application of the Jacobian integral is to establish regions of possible motion. The Jacobian integral may be written as

$$v^2 = 2\Omega(x,y) - C.$$

Let the speed relative to the synodic coordinate system by $v_0$ at a given point $(x_0, y_0)$ along the orbit. The value of the Jacobian constant on this orbit is

$$C_0 = 2\Omega(x_0, y_0) - v_0^2.$$

Since this value is the same for any point $(x,y)$ on the orbit, we have

$$C_0 = 2\Omega(x,y) - v^2,$$

where $v$ is the speed at the point $P(x,y)$. Solving the above equation for $v^2$, we have

$$v^2 = 2\Omega(x,y) - C_0.$$

Selecting now a point, $P(x,y)$ so that $2\Omega(x,y)$ is larger than the computed value of $C_0$, then $v^2 > 0$ and motion is possible at such a point. If, on the other hand

$$2\Omega(x,y) \leq C_0.$$

then $v^2 \leq 0$ and motion cannot take place at the point $P(x,y)$.

The regions of possible motion are, therefore, separated from the forbidden regions by curves which are given by the equation

$$2\Omega(x,y) = C_0.$$

Along these curves the speed is zero and for this reason these curves are called curves of zero velocity or Hill's curves.

To establish the regions of possible motion, the curves $\Omega =$ constant must be constructed. In what follows these "equipotential" or niveau curves will be discussed.

According to the value of $C$ there are several different cases representing different problems as well as physical situations. To review these cases we start with $C \to \infty$ and decrease its value to $C = 3$.

Case I. Large values of $C$ imply large values of $\Omega$.

This occurs if one of the following conditions is approached:

$$r_1 \to 0, r_2 \to 0 \quad \text{or} \quad r_1 \to \infty \text{ and } r_2 \to \infty.$$

So the curves of zero velocity are approximately circles arount $P_1$ and $P_2$. These ovals shrink as C increases, so if a zero velocity oval corresponding to a given $C_1$ value is constructed around $P_1$, another zero velocity oval with $C = C_2$ will be inside of the first one if $C_2 > C_1$. If a particle has a given velocity and position vector, i.e. its state of motion is defined giving for its Jacobian constant a value equal to $C_1$ then it can move only inside of the $C_1$ oval. This follows if we consider that one the $C_1$ oval

$$v^2 = 2\Omega(x, y) - C_1 = 0$$

and inside of this oval there are $2\Omega(x,y) - C_2 = 0$ curves for various C values with $C_2 > C_1$. Now if the particle has $C_1$ as its Jacobian constant and it is located on a $2\Omega(x,y) = C_2$ curve, then the square of its velocity must be larger than zero on this curve, since $C_2 > C_1$. The particle with $C_1$ can not go outside the $C_1 = 2\Omega$ oval since it would encounter points for which the zero velocity oval is associated with $C_3 < C_1$ which would require a decrease in the particle's $v^2$ from zero, i.e. it would require imaginary velocity.

In fact it is generally true that motion always takes place on that side of the zero velocity surface where the constant $\Omega$ lines have positive gradients, i.e. where C is increasing.

The condition $r_1 \to \infty$ (and consequently $r_2 \to \infty$) also results in large $\Omega$ and so in large C values. In this case $r_1 \simeq r_2 \simeq r$ and $\Omega \cong r^2/2$ with $v^2 = r^2 - C$

The zero velocity ovals can be approximated by circles with radii

$$r_0 = C.$$

These circles expand as C increases, therefore motion is possible outside the curve of zero velocity. If a particle moves outside such a zero velocity oval, it will never cross it, i.e. the particle will never change its planetoid or comet characteristics and it will never become a satellite.

As C is decreased, the two ovals described above will increase in size and will touch at $L_2$. At the same time the large outside oval shrinks. As C is decreased further the two inside ovals unite permitting motion inside of the curve. In this case therefore satellite exchange might take place since particles are allowed to wander from the neighborhood of $m_1$ to the neighborhood of $m_2$. The practical significance of this case in connection with earth-lunar trajectories is apparent. It is noted, nevertheless, that an interchange of particles between the inside region of the $C_2$ curve enclosing $m_1$ and $m_2$ and the outside region of the large oval (corresponding to the same $C_2$) is still not permissible.

# Curves of Zero Velocity

Further decrease of C will cause the large outside oval and the inside figure to become in contact at $L_3$ or at $L_1$ depending on the value of $\mu$. (The contact point is established on the side of the smaller mass).

As C is decreased further the figure opens up at $L_1$ allowing a communication between the external and internal areas for the first time. The next step is when the central portion of the horseshoe like forbidden area narrows and its width becomes zero intersecting the $x$ axis at $L_3$.

As C is further decreased, the curve passing through $L_3$ separates at $L_3$ and the two areas (one above and the other one below the $x$ axis) start shrinking toward $L_4$ and $L_5$. When C reaches its minimum value (C = 3) the forbidden areas shrink to zero. When $C \leq 3$, motion is possible everywhere.

A summary of the regions of motion is shown on Figure 4 for various C values. The figure which is correct only in a topological sense was constructed for $\mu < 1/2$

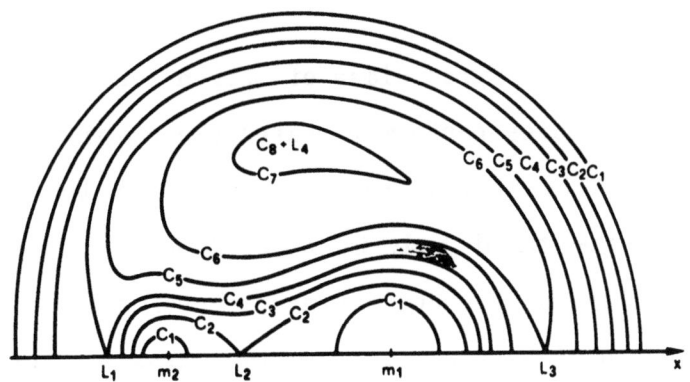

Fig. 4. Summary of the Hill Curves for $\mu < 1/2$, $C_1 > C_2 > \cdots > C_8 = 3$.

## The General Problem of Three Bodies

The basic parameter is the total energy of the system, which here we denote by $h$. The case of <u>positive total energy</u> may be disposed of quickly since it leads to disruption of the system. Either all participating bodies depart on hyperbolic orbits according to $|\bar{r}_{ij}| \to t$, a motion which may be termed <u>explosion</u>, or two of the bodies form a binary, $|\bar{r}_{12}| \to a$, and the third body increases its distance from this binary according to the hyperbolic low : $|\bar{r}_{13}|$ , $|\bar{r}_{22}| \to t$. This motion is called hyperbolic-elliptic by Chazy and will be referred to here as <u>escape</u>. These unbounded motions correspond to results known from the behaviour of two bodies where, when $h > 0$, the motion is unbounded (hyperbolic). This similarity between the two and three-body problems, however, is not complete since $h < 0$ does not, in general, correspond to the behaviour of the problem of three bodies.

The first class of motion for $h > 0$ is called <u>interplay</u>. The bodies perform repeated close approaches and $|\bar{r}_{ij}| < a$. Another class is termed <u>ejection</u>, when two bodies form a binary while the third body is ejected with elliptic relative velocity. Chazy's "bounded motion", therefore, may be separated in the modern classification into two classes : interplay and ejection. As the energy of the ejected body increases it may depart on a hyperbolic orbit, leaving the binary behind. This unbounded motion occurring with $h < 0$ is called (as before for $h > 0$) escape or "hyperbolic-elliptic" by Chazy. An important special case for applications of bounded motions is termed revolution when the binary formed is surrounded by the orbit of the third body. This motion occurs only with $h < 0$ and its stability depends on the magnitude of the ratio $\rho/r$ where $\rho$ is the distance between the center of mass of the binar and the third body, and $r$ is the distance between the members of the binary. If this ratio is large, the system is stable but the original definitions of the general problem of three bodies is violated since this motion, of course, changes into interplay if $\rho/r \simeq 1$.

Another special case of bounded motion is termed equilibrium configurations consisting of the triangular and collinear Lagrangean solutions. It is known that these solutions are unstable when the masses are of the same order of magnitude and consequently they transit into interplay.

Finally periodic orbits must be mentioned which are also bounded and as far as known, unstable. The periodic orbits of the general problem do not seem to form families in the same sense we know families in the restricted problems and they are not dense.

Note that to this classification one should add the case of $h = 0$ as well as the parabolic behaviors. Chazy's hyperbolic-elliptic or parabolic motion ($|\bar{r}_{ij}| \longrightarrow t^{2/3}$) for $h = 0$

are of limited significance because they call for a specific value of the energy constant. His hyperbolic-parabolic motion ($|\bar{r}_{12}| \to t^{2/3}$; $|\bar{r}_{13}|$, $|\bar{r}_{23}| \to t$) occurring with $h > 0$ and parabolic-elliptic behavior ($|r_{12}| < a$; $|r_{13}|$, $|r_{23}| \to t^{2/3}$) for $h < 0$ are also of lower dimensionality. These classes, of course, separate the corresponding hyperbolic and elliptic cases.

Birkhoff's classification is based on the moment of inertia (I) of the three bodies, i.e. on the behavior of the function $I(t)$. In this classification <u>escape</u> for $h < 0$ is associated with $I \to \infty$. <u>Interplay</u> and <u>periodic orbits</u> corresponds to a uniformly bounded behavior of $I(t)$. The <u>equilibrium solutions</u> correspond to $I = $ constant. With an oscillatory behavior of $I(t)$ - such that one of the three bodies recedes arbitrarily far and returns - we may associate an extreme case of <u>ejection</u>. Such oscillatory motion can occur, of course, only for $h < 0$.

One of the important results of the past five years in research on the three-body problem is the discovery that the class of motions termed <u>escape</u> dominates (Agekyan 1967; Szebehely 1967). In other words, for arbitrary initial conditions and after a sufficiently long time the outcome of the motion is hyperbolic-elliptic. This result verifies an "opinion" of Birkhoff according to which it is "possible that the motions for which $I \to \infty$ as $t \to \infty$ fill up the manifold of possible motions densely". If we add to this the conjectures that periodic orbits are neither densely distributed nor are they stable in

# Analysis of Escapes

the general problem, then the basic system of solutions of the three-body problem seems to be the escape type. Such solutions are unstable according to Laplace's definition of stability but show a remarkable persistence to changes in the initial conditions. To establish families of periodic orbits according to what is known today, requires changes in the participating masses as well as in the initial conditions; consequently they do not seem to be densely distributed (Standish 1970; Szebehely 1967, 1970). On the other hand the escape type orbits form continuous families as the initial conditions are changed (Szebehely 1973). For $|h|$ small, one might expect that the manifold of motion will be filled with escape orbits since this is the case for $h > 0$. As long as no families of stable periodic orbits exist, the conjecture of densely distributed escape orbits is feasible. Numerical results seem to indicate this since the various types of motions described above all have the tendency to turn into escape orbits. Interplay is the necessary prelude leading to escape or ejection. Repeated ejections turn into escapes. Solutions near the Lagrangean solutions are unstable for the general case, as mentioned before, and turn into interplays. The known unstable periodic orbits are also surrounded by interplays as are revolutions unless restrictions are imposed in the distances.

The simplified model which explains approximately the behavior of the system consists of an already formed binary with bounding energy $E_b = -Gm_1m_2/2a$ and of another two-body problem formed by $(m_1 + m_2)$ and $m_3$. The energy of this system is $E_e$ and may be obtained by

$$E_e = \frac{m_3}{2} v_3^2 + \frac{m_1 + m_2}{2} v_{12}^2 - \frac{(m_1 + m_2) m_3 G}{d},$$

where $m_1$ and $m_2$ are the masses of the binary with semi-major axis $a$, $m_3$ is the mass of the ejected or escaping body with velocity $v_3$, $d$ is the distance between the center of mass of $m_1 + m_2$ and $m_3$, and $v_{12}$ is the velocity of the center of mass of the binary.

The total energy of the system is $h = E_t = E_b + E_e$. If $E_t > 0$ then, since $E_b < 0$ always, we have that $E_e \geq 0$, and we conclude that with positive or zero total energy binary formation gives escape. This means that for $E_t > 0$ there is no ejection. The above result will be shown later to be exact.

If $E_t < 0$ the situation is more complicated. Since $E_e = E_t - E_b = |E_b| - |E_t|$, for escape $|E_b| > |E_t|$ is required. This can always be established with sufficiently small value of $a$; Therefore escape for negative total energy is associated with the formation of a close binary. If the binary formed does not have enough negative energy because its semi-major axis is large, escape does not occur but it is replaced by an ejection.

Jacobian Coordinates

The introduction of the variables proposed by Lagrange and by Jacobi is identical with utilizing the center of

# Jacobian Coordinates

mass integrals for reducing the order to the system to 12 from 18. On Fig. 5 the Jacobian vectors are $\bar{r}$ and $\bar{\rho}$, the first connecting $m_1$ and $m_2$ and the second the center of mass of $m_1$ and $m_2$ with $m_3$. In order to transform the equations of motion into the Jacobian system we express $\bar{r}_{21}$, $\bar{r}_{32}$ and $\bar{r}_{31}$ with $\bar{r}$ and $\bar{\rho}$. The result is

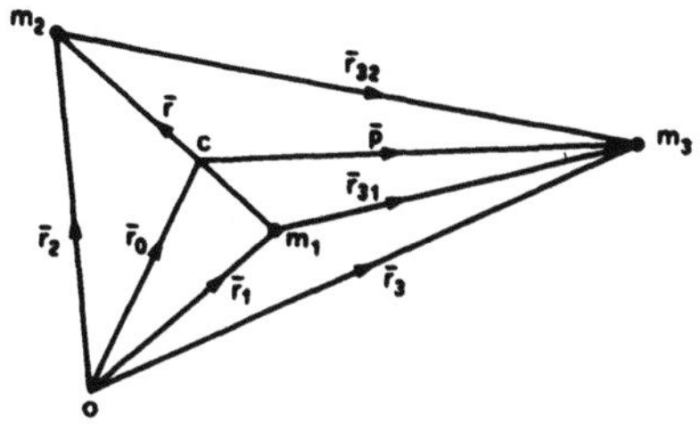

Fig. 5. Jacobian Coordinates

$$\bar{r}_{21} = \bar{r}_2 - \bar{r}_1 = \bar{r},$$

$$\bar{r}_{32} = \bar{r}_3 - \bar{r}_2 = \bar{\rho} - \frac{m_1}{\mu} \bar{r},$$

$$\bar{r}_{31} = \bar{r}_3 - \bar{r}_1 = \bar{\rho} + \frac{m_2}{\mu} \bar{r},$$

since the vector pointing from $m_1$ to C is $(m_2/\mu)\bar{r}$ and from C to $m_2$ is $(m_1/\mu)\bar{r}$ with $\mu = m_1 + m_2$.

Note also that $\bar{r}_3 = \mu/M\,\bar{\rho}$, since $\bar{r}_3 = \bar{r}_0 + \bar{\rho}$ and $r_0\mu + r_3 m_3 = 0$ or $r_0 = -r_3 m_3/\mu$, having O at the center of mass of the system.

Now the three equations of motion are ready to be transformed. From the equations of motion we have

$$\ddot{\bar{r}}_1 = G \frac{m_2}{r_{12}^3} \bar{r}_{21} + G \frac{m_3}{r_{31}^3} \bar{r}_{31},$$

$$\ddot{\bar{r}}_2 = G \frac{m_3}{r_{23}^3} \bar{r}_{32} + G \frac{m_1}{r_{12}^3} \bar{r}_{12},$$

$$\ddot{\bar{r}}_3 = G \frac{m_1}{r_{13}^3} \bar{r}_{13} + G \frac{m_2}{r_{23}^3} \bar{r}_{23}$$

Subtracting the first equation from the second we have

$$\ddot{\bar{r}} = -G\mu \frac{\bar{r}}{r^3} + Gm_3 \left( \frac{\bar{r}_{32}}{r_{32}^3} - \frac{\bar{r}_{31}}{r_{31}^3} \right),$$

or

$$\ddot{\bar{r}} = -G\mu \frac{\bar{r}}{r^3} + Gm_3 \left( \frac{\bar{\rho} - \frac{m_1}{\mu}\bar{r}}{r_{32}^3} - \frac{\bar{\rho} + \frac{m_2}{\mu}\bar{r}}{r_{31}^3} \right).$$

A substitution into the third equation of motion gives

$$\ddot{\bar{\rho}} = -\frac{M}{\mu} G \left( \frac{m_1 \bar{r}_{31}}{r_{31}^3} + \frac{m_3 \bar{r}_{32}}{r_{32}^3} \right),$$

or

$$\ddot{\bar{\rho}} = -\frac{M}{\mu} G \left[ \frac{m_1 \left( \bar{\rho} + \frac{m_2}{\mu}\bar{r} \right)}{r_{31}^3} + \frac{m_2 \left( \bar{\rho} - \frac{m_1}{\mu}\bar{r} \right)}{r_{32}^3} \right]$$

The last equations for $\ddot{\bar{r}}$ and for $\ddot{\bar{\rho}}$ form a 12-th order system using Jacobian coordinates. Two remarks are in order. First note that these equations may also be written in the short and elegant form :

$$\ddot{\bar{r}} + \mu \bar{f}(\bar{r}) = (M - \mu) \left[ \bar{f}(\bar{\rho} - \nu \bar{r}) - \bar{f}(\bar{\rho} + \nu * \bar{r}) \right]$$

# Jacobian Coordinates

and

$$\ddot{\bar{\rho}} = -M\left[\nu^* \bar{F}(\bar{\rho} - \nu\bar{r}) + \nu\bar{F}(\bar{\rho} + \nu^*\bar{r})\right],$$

where $F(x) = G\, x|x|^{-3}$, $\nu = \dfrac{m_1}{\mu}$ and $\nu^* = \dfrac{m_2}{\mu}$.

The second note is more interesting. Since

$$F = G\left(\frac{m_1 m_2}{r} + \frac{m_2 m_3}{r_{23}} + \frac{m_3 m_1}{r_{31}}\right) = F(\bar{r}, \bar{\rho})$$

we may obtain with some diligence that

$$\ddot{\bar{r}} = \frac{1}{g_1}\frac{\partial F}{\partial \bar{r}}$$

and

$$\ddot{\bar{\rho}} = \frac{1}{g_2}\frac{\partial F}{\partial \bar{\rho}},$$

where

$$g_1 = \frac{m_1 m_2}{\mu} \quad \text{and} \quad g_2 = \frac{m_3 \mu}{M}.$$

From these an integral of energy follows immediately since

$$g_1 \dot{\bar{r}}\ddot{\bar{r}} + g_2 \dot{\bar{\rho}}\ddot{\bar{\rho}} = \frac{\partial F}{\partial r}\dot{\bar{r}} + \frac{\partial F}{\partial \rho}\dot{\bar{\rho}}$$

or

$$\frac{1}{2}\left(g_1 \dot{\bar{r}}^2 + g_2 \dot{\bar{\rho}}^2\right) = F + h.$$

Note that in fact $T = \dfrac{1}{2}\left(g_1 \dot{\bar{r}}^2 + g_2 \dot{\bar{\rho}}^2\right)$ as may be shown by direct substitutions. Similarly, the angular momentum may be written as

$$\bar{c} = g_1 \bar{r} \times \dot{\bar{r}} + g_2 \bar{\rho} \times \dot{\bar{\rho}} \;.$$

As an exercise we may show that

$$\dot{\bar{c}} = g_1 \bar{r} \times \ddot{\bar{r}} + g_2 \bar{\rho} \times \ddot{\bar{\rho}} = 0 \;,$$

since

$$g_1 \bar{r} \times \ddot{\bar{r}} = - g_2 \bar{\rho} \times \ddot{\bar{\rho}} = \frac{G}{\mu} m_1 m_2 m_3 \, \bar{r} \times \bar{\rho} \left( \frac{1}{r_{32}^3} - \frac{1}{r_{31}^3} \right) \;.$$

At this point we introduce the moment of inertia of the three bodies $I$, which will play an important role in the sequence. In general, the moment of inertia with respect to the origin of the coordinate system is defined by

$$I = \sum_{i=1}^{3} m_i \bar{r}_i^2 \;.$$

It may be shown that the moment of inertia with respect to the center of mass is

$$\Phi = \sum_{1 \leq i < j \leq 3} \frac{m_i m_j}{M} \bar{r}_{ij}^2 \;,$$

which expression is also known as the Jacobian functions.

In the following text the expressions for $I$ and $\Phi$ will be used alternatively and since the origin of the coordinate system will be at the center of mass, we have $I = \Phi$.

# Lagrange–Jacobi Equation

Using the Jacobian coordinates

$$I = g_1 \bar{r}^2 + g_2 \bar{\rho}^2 ,$$

which may be shown by substituting the Jacobian transformation into the Jacobian function.

**The Lagrange-Jacobi Equation**

This equation, basic in the science of stellar dynamics, was first given by Lagrange for the problem of three bodies in 1772.

The equation may be written as

$$\ddot{I} = 2(2T - F),$$

which may be transformed to various forms by $h = T - F$, such as

$$\ddot{I} = 2(F + 2h) \quad \text{and} \quad \ddot{I} = 2(T + h).$$

Using the Jacobian system the proof is as follows. From

$$I = g_1 \bar{r}^2 + g_2 \bar{\rho}^2 ,$$

we have

$$\ddot{I} = 2(g_1 \dot{\bar{r}}^2 + g_2 \dot{\bar{\rho}}^2 + g_1 \bar{r}\ddot{\bar{r}} + g_2 \bar{\rho}\ddot{\bar{\rho}}),$$

where the first two terms represent twice the kinetic energy. The last two terms become

$$\bar{r} \frac{\partial F}{\partial \bar{r}} + \rho \frac{\partial F}{\partial \bar{\rho}} = -F,$$

where the left side is obtained by substituting the equations of motion and the right side results form Euler's theorem of homogeneous functions applied to F of order -1. In this way the proof is complete.

From the Lagrange-Jacobi equation it follows that for $h > 0$, $\ddot{I} \geq 4h \geq 0$ and $I \geq 2h t^2 + bt + c$. Therefore $I \to \infty$ as $t \to \infty$ and at least one of the distances $r_{ij} \to \infty$. The same may be shown for $h = 0$. Furthermore, since $I = g_1 r^{-2} + g_2 \rho^{-2}$, as $I \to \infty$, $\rho \to \infty$, if $r$ is bounded. If a binary is formed, $|r| < a$, consequently as $I \to \infty$, $\rho \to \infty$, which corresponds to escape. If no binary is formed, $(r, \rho) \to \infty$ as $I \to \infty$, which corresponds to explosion.

For a system with positive total energy $\ddot{I} \geq 2h > 0$, consequently the curve $I(t)$ is convex (from below). If a system with $h > 0$ begins its motion at $t = 0$, we have $I(0) > 0$, $\dot{I}(0) \gtreqless 0$ and $I_0 > 0$. Note that $\dot{I}(0) = 2[g_1 \bar{r}(0) \dot{\bar{r}}(0) + g_2 \bar{\rho}(0) \dot{\bar{\rho}}(0)]$ may be positive or negative and, in fact, even zero without the initial velocities cannot be zero). The same applies for $h = 0$.

If the total energy is negative, we have initially that $I(0) > 0$, $\dot{I}(0) \gtreqless 0$ and $\ddot{I}(0) \gtreqless 0$. Consider first a system with zero initial velocities. In this case $\dot{I}(0) = 0$, $T(0) = 0$, $\ddot{I}(0) = 2h < 0$. The curve $I(t)$ is concave from below initially

# Condition for Escape

and a contraction takes place with $\ddot{I} < 0$. At the same time F increases (and so does T) and when the value of $F + 2h = 0$, or $F = = +2|h|$ is reached, $\ddot{I}$ becomes zero. After this $\ddot{I} > 0$ and the curve is convex from below. Now the opposite trend takes place and the contraction after reaching a value $I_{min}$ with $\dot{I} = 0$ turns into an expansion with $\dot{I} > 0$ and still with $\ddot{I} > 0$. As the expansion decreases the value of F, the quantity $F + 2h$, becomes zero and then negative again and the curve $I(t)$ will be concave once again from below. Now let us consider once more a contraction with $\dot{I} > 0$ without restricting $\dot{I}(0)$ to 0. At the minimum of I the value of F is large and the bodies are close together. After this time F decreases and the value of $\ddot{I}$ stays positive as long as $F > 2|h|$. Therefore an explosion, when all $r_{ij}$ increase their values to infinity, is impossible for $h > 0$ since when $r_{ij} \to \infty$, $F \to 0$, In this process F will reach the value of $2|h|$ at which time $\ddot{I}$ becomes negative. The only way F can stay larger than $2|h|$ for all time is if a binary is formed.

Let the masses of the members of the binary be $m_1$ and $m_2$. Then the condition for $\ddot{I} > 0$ is

$$F = G\left(\frac{m_1 m_2}{r} + \frac{m_2 m_3}{r_{23}} + \frac{m_3 m_1}{r_{31}}\right) > 2|h|.$$

With sufficiently small value of r this condition may be satisfied, no matter how large the other distances become. The condition is satisfied if, as F varies,

$$F_{min} > 2|h|.$$

Now as $\rho \to \infty$, $r_{23}$ and $r_{31} \to \infty$ and

$$F_{min} = G\frac{m_1 m_2}{r_{max}},$$

where $r_{max} = a(1 + e)$ is the apogee distance of the primary in its asymptotic state. The escape condition now becomes

$$G\frac{m_1 m_2}{a(1+e)} > 2|h|$$

$$\frac{G m_1 m_2}{2|h|(1+e)} > a.$$

Note that if a circular binary orbit is formed ($e_1 = 0$),

$$a_1 < \frac{G m_1 m_2}{2|h|}$$

and if the eccentricity is high ($e_2 = 1$),

$$a_2 < \frac{a_1}{2},$$

therefore highly eccentric binary orbits require close binaries.

Sundman's Inequality

This important result connects the magnitude of the angular momentum vector $|\bar{c}| = c$ with the moment of inertia by the following inequality:

# Sundman's Inequality

$$c^2 < 2IT - \frac{1}{4}\dot{I}^2.$$

The conventional proof uses the triangle and Cauchy's inequalities. The magnitude of the angular momentum is

$$c = \left|\sum m_i \vec{r}_i \times \vec{v}_i\right| < \sum m_i r_i v_i \left|\sin \alpha_i\right| = \sqrt{m_i}\, r_i \sqrt{m_i}\, v_i \left|\sin \alpha_i\right|,$$

where $\alpha_i$ is the angle between $\vec{r}_i$ and $\dot{\vec{r}}_i = \vec{v}_i$.

From this, by Cauchy's inequality we have

$$c^2 \leq \sum m_i r_i^2 \sum m_i v_i^2 \sin^2 \alpha_i.$$

On the other hand

$$\dot{I} = 2\sum m_i \vec{r}_i \dot{\vec{r}}_i = 2\sum m_i r_i v_i \cos \alpha_i = 2\sum \sqrt{m_i}\, r_i \sqrt{m_i}\, v_i \cos \alpha_i$$

from which once again by Cauchy's inequality, we have

$$\frac{1}{4}\dot{I}^2 < \sum m_i r_i^2 \sum m_i v_i^2 \cos^2 \alpha_i,$$

Addition of the last two inequalities gives the desired result.

An important variation of Sundman's inequality is obtained when the kinetic energy is eliminated by means of the Lagrange-Jacobi equation:

$$c^2 \leq (\ddot{I} - 2h)I - \frac{1}{4}\dot{I}^2.$$

Note that a weaker form of this is

$$c^2 \leq (\ddot{I} - 2h)I$$

which is often used.

In the following another, rather useful form of Sundman's inequality will be derived. Dividing by $I > 0$, rearranging and multiplying by $2/\sqrt{I}$ gives

$$0 \le (\ddot{I} - 2h - \frac{c^2}{I^2} - \frac{\dot{I}^2}{4I^2}) \frac{2}{\sqrt{I}} = Z.$$

In this way an integrable combination is produced for the non-negative function $Z$. In gact it may be shown that

$$\frac{dL}{dt} = Z \frac{dI}{dt},$$

where

$$L = \frac{1}{\sqrt{I}} (\dot{I}^2 + 4c^2) - 8h \sqrt{I}.$$

The proof only requires the computation of the time-derivative of $L$ and then obtaining $Z$ as $\dot{L}/\dot{I}$.

This form of the inequality says that since $Z > 0$, if $I$ increases $L$ does not decrease or if $I$ decreases $L$ does not increase.

Consider the simplest form of Sundman's inequality

$$c^2 \le (\ddot{I} - 2h) I.$$

Rearrangement gives

$$\frac{c^2}{I} + 2h \le \ddot{I} \quad \text{or} \quad \frac{c^2}{2|h|} - I \le \frac{\ddot{I} I}{2|h|},$$

for $h > 0$. The sign of $\ddot{I}$ is controlled by the value of $I$ and introducing the critical value of the moment of inertia,

$$\frac{c^2}{2|h|} = I_c ,$$

we have

$$I_c - I \le \frac{I}{2|h|} \ddot{I} .$$

So as long as $I > I_c$, $\ddot{I}$ is positive.
If, on the other hand, $I < I_c$, $\ddot{I}$ is larger than a negative number, consequently nothing may be said regarding its sign.

Considering the complete inequality, we have

$$I_c - I + \frac{\dot{I}^2}{8|h|} \le \frac{\ddot{I} I}{2|h|}$$

consequently as long as

$$I \le I_c + \frac{\dot{I}^2}{2|h|} ,$$

we have $\ddot{I} > 0$. Before an inflexion point is reached, $I$ certainly will be higher than $I_c$; how much higher depends on $\dot{I}$. A fast expansion with high value of $\dot{I}$ will carry $I$ higher without inflexion than a slow expansion.

Returning now to the simplified inequality we consider that part of the curve $I(t)$ for which $I \le I_c$ and have $\dot{I} = 0$ at $I_1$. This will be a proper minimum with $\ddot{I} > 0$. The curve will rise on both sides of $I_1$ untill another point is reached, say $I$, where $I > 0$. From Sundman's modified inequality we have that if $I_{min} = I_1 < I_2$, corresponding to $\dot{I}_1 = \dot{I}_2 = 0$ then $L_1 \leq L_2$ and with $h < 0$

$$\frac{4c^2}{\sqrt{I_1}} + 8|h|\sqrt{I_1} \leq \frac{4c^2}{\sqrt{I_2}} + 8|h|\sqrt{I_2}$$

or

$$4c^2 \frac{\sqrt{I_2} - \sqrt{I_1}}{\sqrt{I_1 \cdot I_2}} \leq 8|h|\left(\sqrt{I_2} - \sqrt{I_1}\right).$$

Since $I_2 > I_1$ we have

$$\frac{1}{I_1}\left(\frac{c^2}{2|h|}\right)^2 = \frac{I_c^2}{I_1} \leq I_2 .$$

Note that as $t \to \infty$, $I$ cannot approach either zero or a constant value $I_0 \neq 0$.

Now attention is directed to the above inequality. For a given $I_1 = I_{min} < I_c$ the curve $I(t)$ rises until it becomes at least as large as $I_c^2/I_1$. Note that $I_1 \leq I_c$, therefore

$$I_c \leq \frac{I_c^2}{I_1} \leq I_2 ,$$

so the curve rise above $I_c$ on both sides of $I_1$, see Fig. 6.

Considering $I_c$ fixed, if $I_1$ is sufficiently small, the curve may rise to an arbitrary high $I_2$ value. The value $I_c^2/I_1 = I_3$ is lower than the maximum and when it is reached,

$I_3 > 0$. We now estimate $\dot{I}$ using one again Sundman's inequality with $I_1 = I_{min}$, $\dot{I}_1 = 0$, $I_2 > 1 > I_c$ and $\dot{I} \neq 0$. Since $L_1 \leq L$, we have

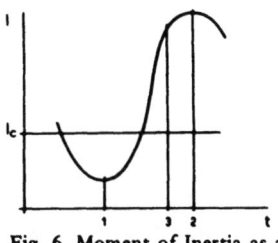

Fig. 6. Moment of Inertia as a Function of Time.

$$\frac{4c^2}{\sqrt{I}} + 8|h|\sqrt{I_1} \leq \frac{4c^2}{\sqrt{I}} + 8|h|\sqrt{I} + \frac{\dot{I}^2}{\sqrt{I}},$$

from which

$$\dot{I}^2 \geq 8|h|\sqrt{I}\left(\sqrt{I} - \sqrt{I_1}\right)\left(\frac{I_c}{\sqrt{I_1}\sqrt{I}} - 1\right).$$

Consequently, for sufficiently low value of $I_1 = I_{min}$, we have $I$ at least as high as $I_c^2/I_1$ with a derivative which is at least as high as given above.

## Bibliography for the Restricted Problem

Birkhoff, G.D.,"The restricted problem of three bodies", Rendiconti de Circolo Matematico di Palermo, Vol. 39, pp. 1-70, (1915).

Birkhoff, G.D.,"Sur le problème restreint des trois corps", Two memoirs, both published in Annali della R. Scuola Normale Superiore di Pisa. The first in series 2, Vol. 4, pp. 267-306 (1935); the second in series 2, Vol. 5, pp. 1-42 (1936).

Birkhoff, G.D.,"Dynamical systems with two degrees of freedom", Trans. Am. Math. Soc., Vol. 18, pp. 199-300 (1919).

Brouwer, D. and G. Clemence, "Methods of Celestial Mechanics", Academic Press, 1961.

Charlier, C.L.,"Die Mechanik des Himmels", Leipzig, von Veit & Co., First Vol. 1902, Second Vol. 1907.

Darwin, G.H.,"Periodic Orbits", Acta Math., Vol. 21, pp. 99-242, (1897).

Goldstein, H.,"Classical Mechanics", Addison-Wesley, (1951).

Hill, G.W.,"Researches in the lunar theory", Am. J. of Math., Vol. 1, pp. 5-26, 129-147, 245-260 (1878).

Klose, A.,"Topologische Dynamik der interplanetaren Massen", Vieteljahrsschrift des Astronomischen Gesellschaft, Vol. 67, pp.61-102, (1932).

Levi-Civita, T., Acta Math., Vol. 42, pp. 99-144 (1919).

Levi-Civita, T.,"Sur la résolution qualitative du problème restreint de trois corps", Acta Mathematica, Vol. 30, pp. 305-327 (1906).

Levi-Civita, T.,"Traiettorie singolari ed urti nel problema ristretto dei tre corpi", Annali di Matematica, Ser. 3, Vol. 9, pp. 1-32, (1904).

Levi-Civita, T., Ann. di Mat., Ser. 3, Vol. 5, pp. 221-309, (1901).

Moulton, F.R., Proc. Math Congr., Cambridge, England, Vol. 2, pp. 182-187 (1913); also, Periodic Orbits, Carnegie Inst., Wash. (1920).

Poincaré, H.,"Les méthodes nouvelles de la mécanique céleste", Paris, 1892, 1893, 1899.

Strömgren, E.,"Connaissance actuelle des orbites dans le problème des trois corps", Bull. Astr. (2), Vol. 9, pp. 87-130 (1935), and Publ. Kbh. Obs. # 100, pp. 1-44 (1935).

Szebehely, V. "Theory of orbits", Academic Press, New York, (1967).

Whittaker E.T., "Analytical dynamics", 4th Edition, Dover, (1944).

Wintner, A., "The analytical foundations of celestial mechanics", Princeton Univ. (1947).

## Bibliography for the General Problem

T.A. Agekian and Zh. P. Anosova, Astronomical Zh., **44**, 1261, (1967).

T.A. Agekian and Zh.P. Anosova, Akad.Arm.SSSR,Astrophys., **4**, 31, (1968).

G.D. Birkhoff, Dynamical Systems, Am.Math.Soc.Publ., Providence, R.I., (1927).

J. Chazy, Bull. Astronomique, **35**, 321 (1918).

J. Chazy, Ann.Sci.Ecole Norm., **39**, 29, (1922).

J. Chazy, Bull.Soc.Math.France, **55**, 222, (1927).

J. Chazy, J.Math.Pure Appl., (9), **8**, 353, (1929).

A. Gautier, Essai historique sur le problème des trois corps, Paris, (1817).

R. Grant, History of Physical Astronomy from the earliest ages to the middle of the nineteenth century, London, (1852).

Y. Hagihara, Celestial Mechanics, MIT Press, Cambridge, Mass., (1970).

R.S. Harrington, Astron. J., **73**, 190, (1968).

C.G.J. Jacobi, Vorlesungen über Dynamik, Reimer Publ., Berlin, (1866).

# Bibliography for the General Problem

E.O. Lovett, Quart. J. Math., 42, 252, (1911).

R. Marcolongo, Il Problema dei Tre Corpi, Hoepli Publ., Milano, (1919).

G.A. Merman, Astron. Zh., 30, 332, (1953).

G.A. Merman, Bull. Inst. Theor. Astr., 6, 69, (1955).

H. Pollard, Mathematical Introduction to Celestial Mechanics, Prentice-Hall Publ., New Jersey, (1966).

H. Pollard, J. of Mathematics and Mechanics, 17, 601, (1967).

H. Pollard and D. Saari, Arch. Rational Mech. and Anal., 30, 263, (1968).

H. Pollard and D. Saari, Celestial Mechanics, 1, 347, (1970).

D. Saari, Trans. Amer. Math. Soc., 162, 267, (1971).

C.L. Siegel, Ann. of Math., (2), 42, 127, (1941).

C.L. Siegel and J. Moser, Lectures on Celestial Mechanics, Springer, Berlin, (1971).

K. Sitnikov, Dokl. Akad. Nauk, USSR, 133, 303, (1960).

E.M. Standish, in Periodic Orbits and Stability, Reidel Publ. Co., p. 375, (1970).

E.M. Standish, Celestial Mechanics, 4, 44, (1971).

E. Stiefel and G. Scheifele, Linear and Regular Celestial Mechanics, Springer Publ., Berlin, (1971).

K.F. Sundman, Acta Math., 36, 105, (1912).

V. Szebehely, Proc. Nat. Acad. of Sciences, USA, 58, 60, (1967).

V. Szebehely, in Periodic Orbits and Stability, Reidel Publ. Co., p. 382, (1970).

V. Szebehely, Celestial Mechanics, 4, 116, (1971).

V. Szebehely, Astronomical Journal, 77, 169, (1972).

V. Szebehely, Celestial Mechanics, 6, 84, (1972).

V. Szebehely, Proc. Nat. Acad. of Sciences, USA, 69, 1077, (1972).

V. Szebehely, Proc. First European Astron.Mtg. in Athens, 1972, Reidel Publ. Co. (1974).

V. Szebehely and F. Peters, Astronomical Journal, 72, 876, (1967).

V. Szebehely and F. Peters, Astronomical Journal, 72, 1187, (1967).

V. Szebehely and T. Feagin, Celestial Mechanics, 8, 11, (1973).

G.A. Tevzadze, Akad. Nauk. Arm. USSR, 15, 67, (1962).

E.T. Whittaker, British Ass. Rept. 1899, p. 121, (1899).

# CONTENTS

| | Page |
|---|---|
| Preface | 3 |
| INTRODUCTION | 5 |
| THE RESTRICTED PROBLEM | 12 |
| Statement of the Problem and Equations of Motion | 12 |
| The Problem of Three Bodies and the Restricted Problem | 16 |
| Regions of Possible Motion | 25 |
| The General Problem of Three Bodies | 30 |
| Jacobian Coordinates | 34 |
| The Lagrange-Jacobi Equation | 39 |
| Sundman's Inequality | 42 |
| Bibliography for the Restricted Problem | 48 |
| Bibliography for the General Problem | 50 |
| Contents | 53 |

MIX
Papier aus verantwortungsvollen Quellen
Paper from responsible sources
FSC® C105338

If you have any concerns about our products,
you can contact us on
**ProductSafety@springernature.com**

In case Publisher is established outside the EU,
the EU authorized representative is:
**Springer Nature Customer Service Center GmbH
Europaplatz 3, 69115 Heidelberg, Germany**

Printed by Libri Plureos GmbH
in Hamburg, Germany